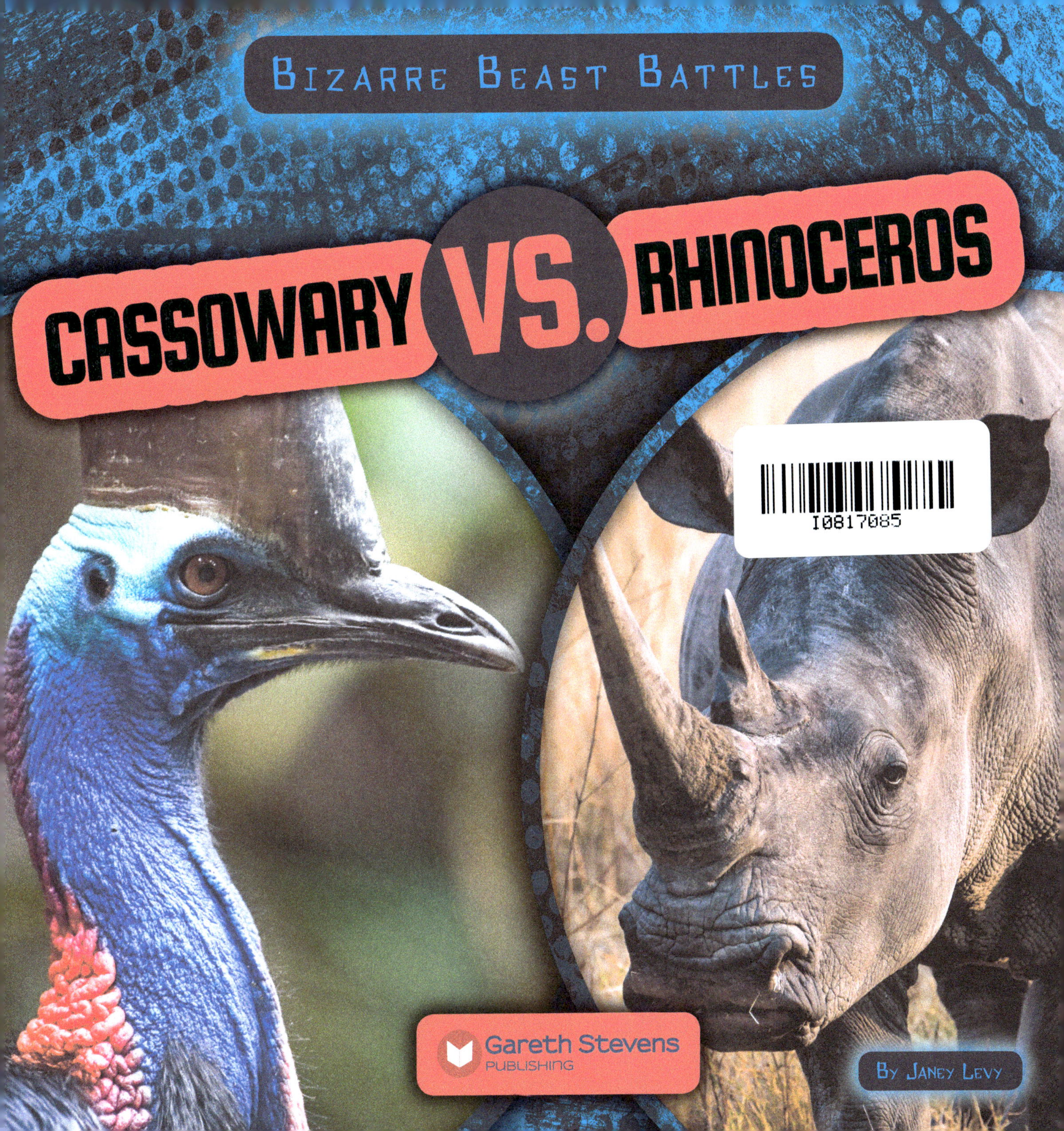
BIZARRE BEAST BATTLES
CASSOWARY VS. RHINOCEROS
I0817085
Gareth Stevens
PUBLISHING
BY JANEY LEVY

Please visit our website, www.garethstevens.com. For a free color catalog of all our high-quality books, call toll free 1-800-542-2595 or fax 1-877-542-2596.

Library of Congress Cataloging-in-Publication Data

Names: Levy, Janey, author.
Title: Cassowary vs. rhinoceros / Janey Levy.
Other titles: Cassowary versus rhinoceros
Description: New York : Gareth Stevens Publishing, [2022] | Series: Bizarre beast battles | Includes index. | Audience: Grades 2–3
Identifiers: LCCN 2020032580 (print) | LCCN 2020032581 (ebook) | ISBN 9781538264713 (library binding) | ISBN 9781538264690 (paperback) | ISBN 9781538264706 (set) | ISBN 9781538264720 (ebook)
Subjects: LCSH: Cassowaries—Juvenile literature. | Rhinoceroses—Juvenile literature.
Classification: LCC QL696.C33 L48 2022 (print) | LCC QL696.C33 (ebook) | DDC 598.5/3—dc23
LC record available at https://lccn.loc.gov/2020032580
LC ebook record available at https://lccn.loc.gov/2020032581

First Edition

Published in 2022 by
Gareth Stevens Publishing
29 E. 21st Street
New York, NY 10010

Designer: Katelyn E. Reynolds
Editor: Therese Shea

Photo credits: Cover, p. 1 (cassowary) Education Images/Universal Images Group via Getty Images; cover, p. 1 (rhino) Michael Beder/Gallo Images/Getty Images Plus; cover, pp. 1–24 (background texture) Apostrophe/Shutterstock.com; pp. 4–21 (cassowary icon) Creative_Outlet/iStock/Getty Images Plus; pp. 4–21 (rhino icon) hypergon/DigitalVision Vectors/Getty Images; pp. 4, 6 MrsWilkins/iStock/Getty Images Plus; p. 5 Australian Scenics/Photolibrary/Getty Images Plus; p. 7 Hannes Thirion/Moment/Getty Images; p. 8 Marc Dozier/The Image Bank/Getty Images; p. 9 Tony Karumba/AFP via Getty Images; p. 10 CraigRJD/iStock/Getty Images Plus; p. 11 Pierre-Yves Babelon/Moment/Getty Images; p. 12 Michele Westmorland/Corbis Documentary/Getty Images; p. 13 Utopia_88/iStock/Getty Images Plus; p. 14 Cees Joppe/500px/Getty Images; p. 15 Mark Chertkow/Gallo Images/Getty Images Plus; p. 16 Philippe Clement/Arterra/Universal Images Group via Getty Images; p. 17 Manoj Shah/Stone/Getty Images; p. 18 mujiri/Shutterstock.com; p. 19 Juan Carlos Munoz/The Image Bank/Getty Images Plus; p. 21 (cassowary) Mark Newman/The Image Bank/Getty Images; p. 21 (rhino) Sarah8000/iStock/Getty Images Plus.

Printed in the United States of America

CPSIA compliance information: Batch #CSGS22: For further information contact Gareth Stevens, New York, New York, at 1-800-542-2595.

CONTENTS

Words in the glossary appear in **bold** type the first time they are used in the text.

KILLER CASSOWARIES

Have you ever heard of cassowaries? They're huge, flightless birds that look a bit like ostriches. Three species, or kinds, exist. The southern cassowary is the largest. It's known as the world's deadliest bird! It lives in the **rainforests** of Australia, New Guinea, and nearby islands.

Cassowary dads are great parents. The mothers lay eggs and then leave. The fathers sit on the eggs until they **hatch**. The fathers take care of the chicks for about 9 months. They keep the chicks safe and teach them to find food.

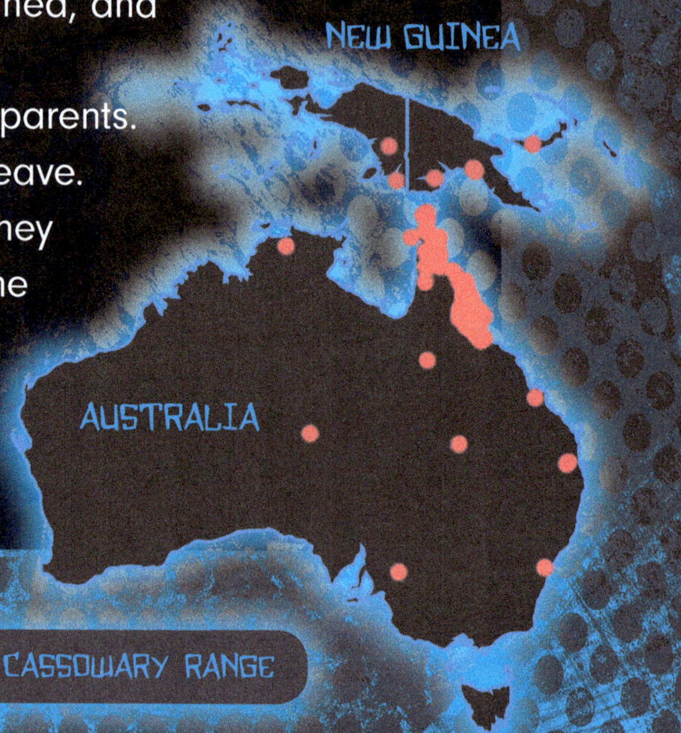

Southern cassowaries have black feathers, a naked blue neck and head, and long red **wattles**. A helmet-like body part called a casque (KASK) sits on top of their head.

BIG, BAD RHINOS

Have you seen a rhinoceros, or rhino, at a zoo? Five species of this giant **mammal** exist. Two live in Africa, and three live in Asia. They make their home in grasslands, wetlands, and forests. The name "rhinoceros" means "nose horn." Some rhinos have two horns, and some have one. The largest are Africa's white rhino and Asia's greater one-horned rhino.

Rhinos are herbivores, or plant eaters. They mostly live alone. However, white rhinos live in groups with around 10 members. When rhinos feel **threatened**, they're dangerous!

White rhinos are actually gray. Their name comes from the Dutch word *weit*, which means "wide."

TAKING THEIR MEASUREMENTS

Cassowaries and rhinos would never actually meet. They live in different parts of the world. But let's see how these two sometimes dangerous animals match up!

Southern cassowaries are almost shockingly large birds. The females are larger than the males. They grow taller and heavier than the average adult human.

CASSOWARY

HEIGHT: UP TO 6 FEET 7 INCHES (2 M)

WEIGHT: UP TO 168 POUNDS (76 KG)

RHINOCEROS
HEIGHT: up to 6 feet (1.8 m)
WEIGHT: up to 6,000 pounds (2,722 kg)

The largest rhinos aren't much shorter. However, they weigh *much* more.

Rhinos weigh about 36 times as much as southern cassowaries. Rhinos win here easily!

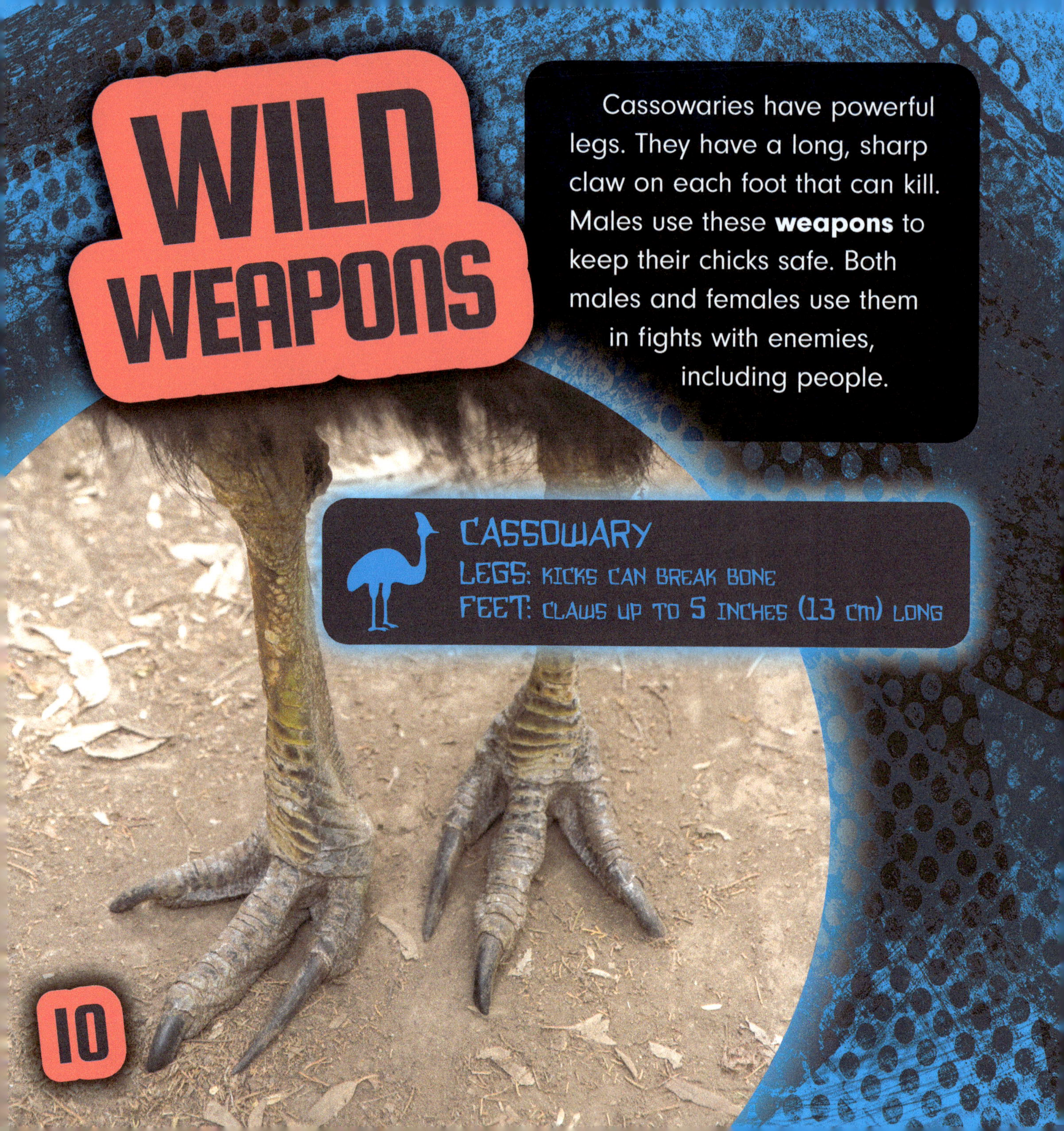

WILD WEAPONS

Cassowaries have powerful legs. They have a long, sharp claw on each foot that can kill. Males use these **weapons** to keep their chicks safe. Both males and females use them in fights with enemies, including people.

CASSOWARY

LEGS: KICKS CAN BREAK BONE

FEET: CLAWS UP TO 5 INCHES (13 cm) LONG

RHINOCEROS

TEETH: UP TO 5 INCHES (13 CM) LONG IN ASIAN RHINOS
HORNS: UP TO 5 FEET (150 CM) LONG IN AFRICAN RHINOS

Asian rhinos have long, sharp teeth that stick out of their mouth. Males use these to fight other males for **mates**. African rhinos lack these teeth and fight with their horns.

Whose weapons are scariest to you?

HIGH SPEED

Like ostriches and other huge birds, cassowaries can't fly. But that doesn't mean they can't travel fast. They race through their forest homes at speeds greater than the fastest human can run!

CASSOWARY
TOP SPEED: 30 MILES (48 KM) PER HOUR

RHINOCEROS
TOP SPEED: 40 MILES (64 KM) PER HOUR

Rhinos may be big and heavy, but that doesn't mean they're slow. They charge at anything they think may be a danger.

Both are fast, but rhinos are faster. They win this round!

ASTONISHING SENSES

Just like you, cassowaries have the senses of hearing, sight, and smell. They use these senses to find food and to **communicate** with each other.

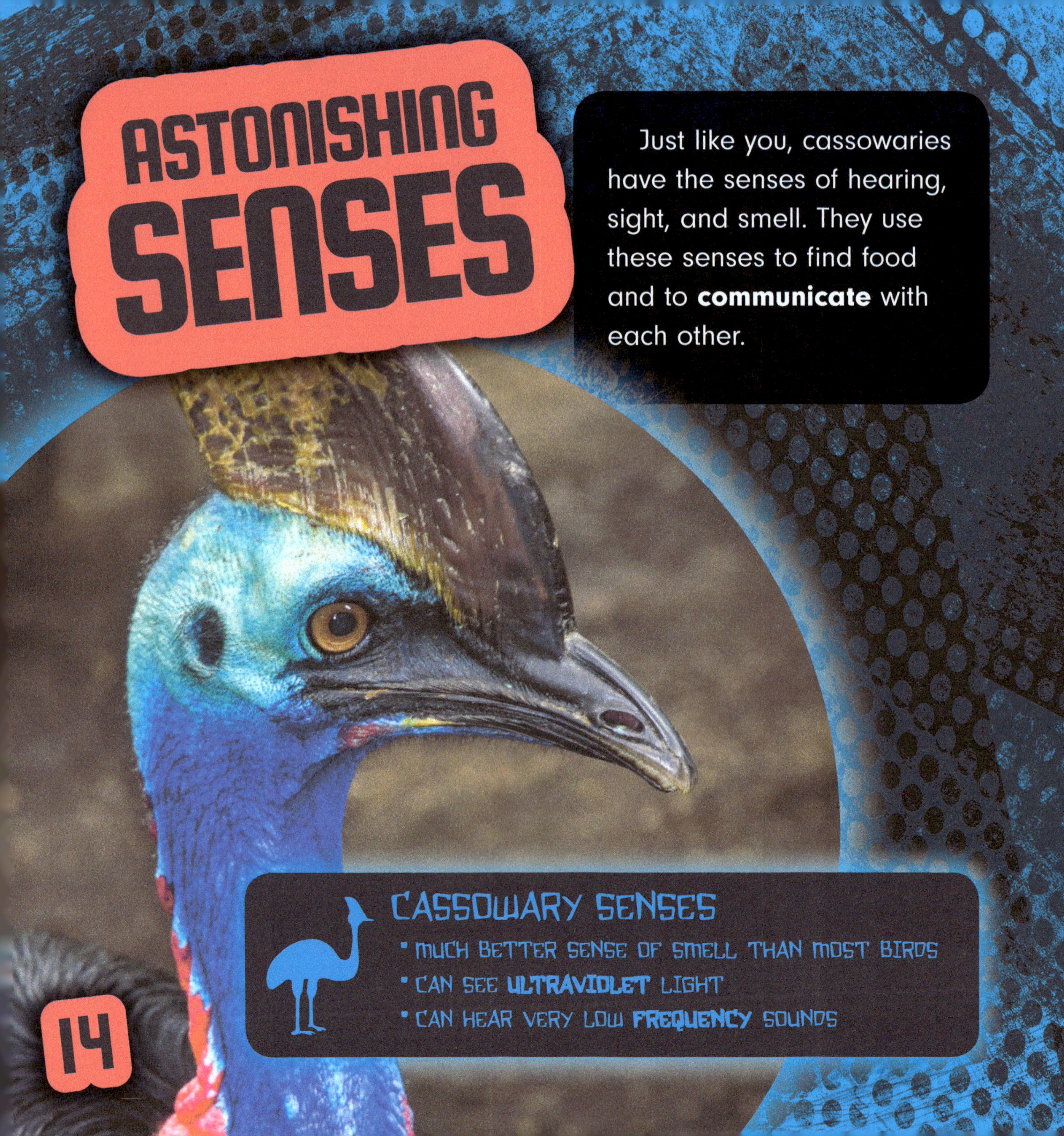

CASSOWARY SENSES

- MUCH BETTER SENSE OF SMELL THAN MOST BIRDS
- CAN SEE **ULTRAVIOLET** LIGHT
- CAN HEAR VERY LOW **FREQUENCY** SOUNDS

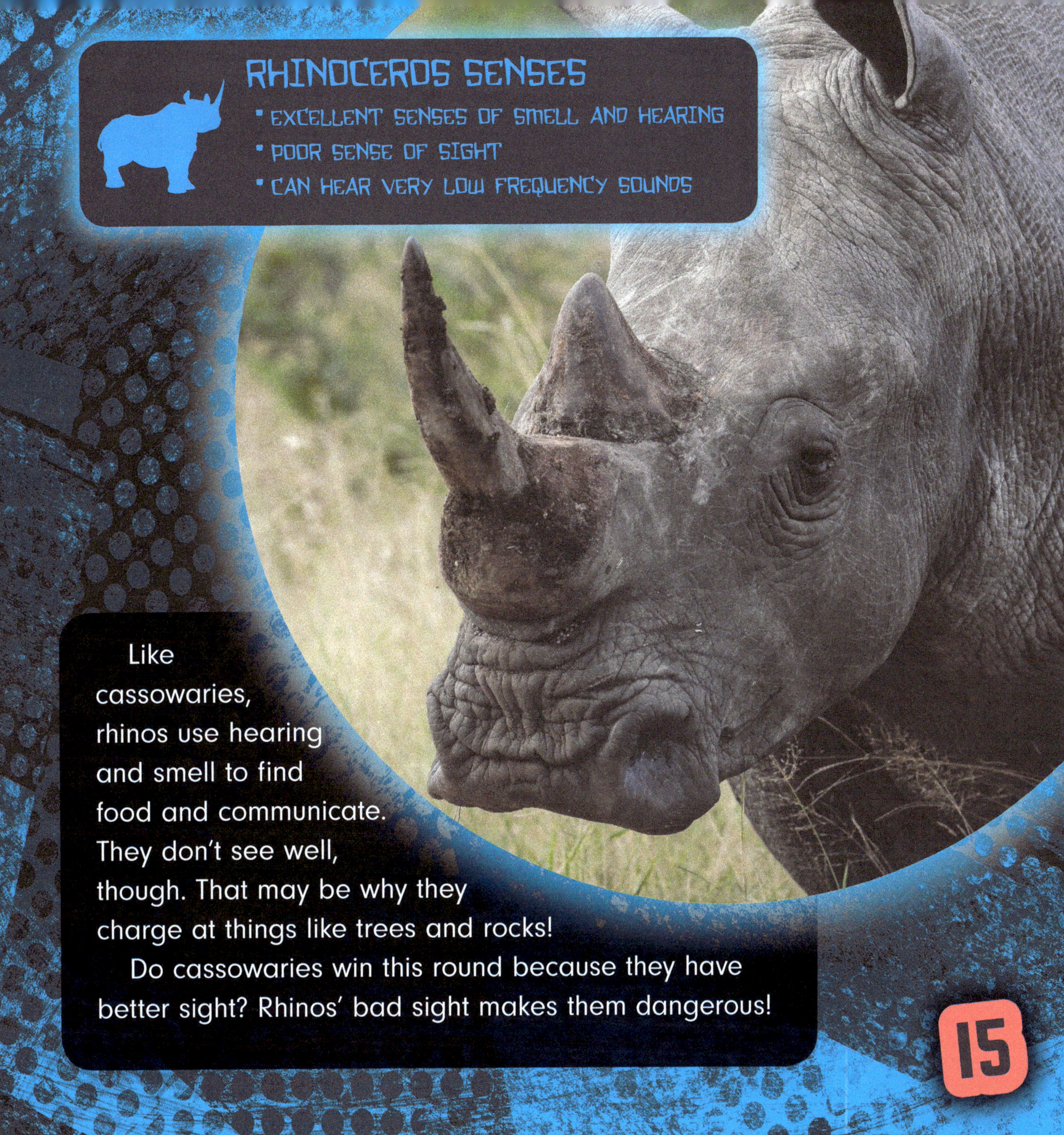

RHINOCEROS SENSES

- EXCELLENT SENSES OF SMELL AND HEARING
- POOR SENSE OF SIGHT
- CAN HEAR VERY LOW FREQUENCY SOUNDS

Like cassowaries, rhinos use hearing and smell to find food and communicate. They don't see well, though. That may be why they charge at things like trees and rocks!

Do cassowaries win this round because they have better sight? Rhinos' bad sight makes them dangerous!

LONG LIFE SPANS

It's hard to know what wild cassowaries' **life span** is because they're shy and not often seen. In zoos, they can live long lives. That's because they're safe from predators and harm to their **ecosystem**.

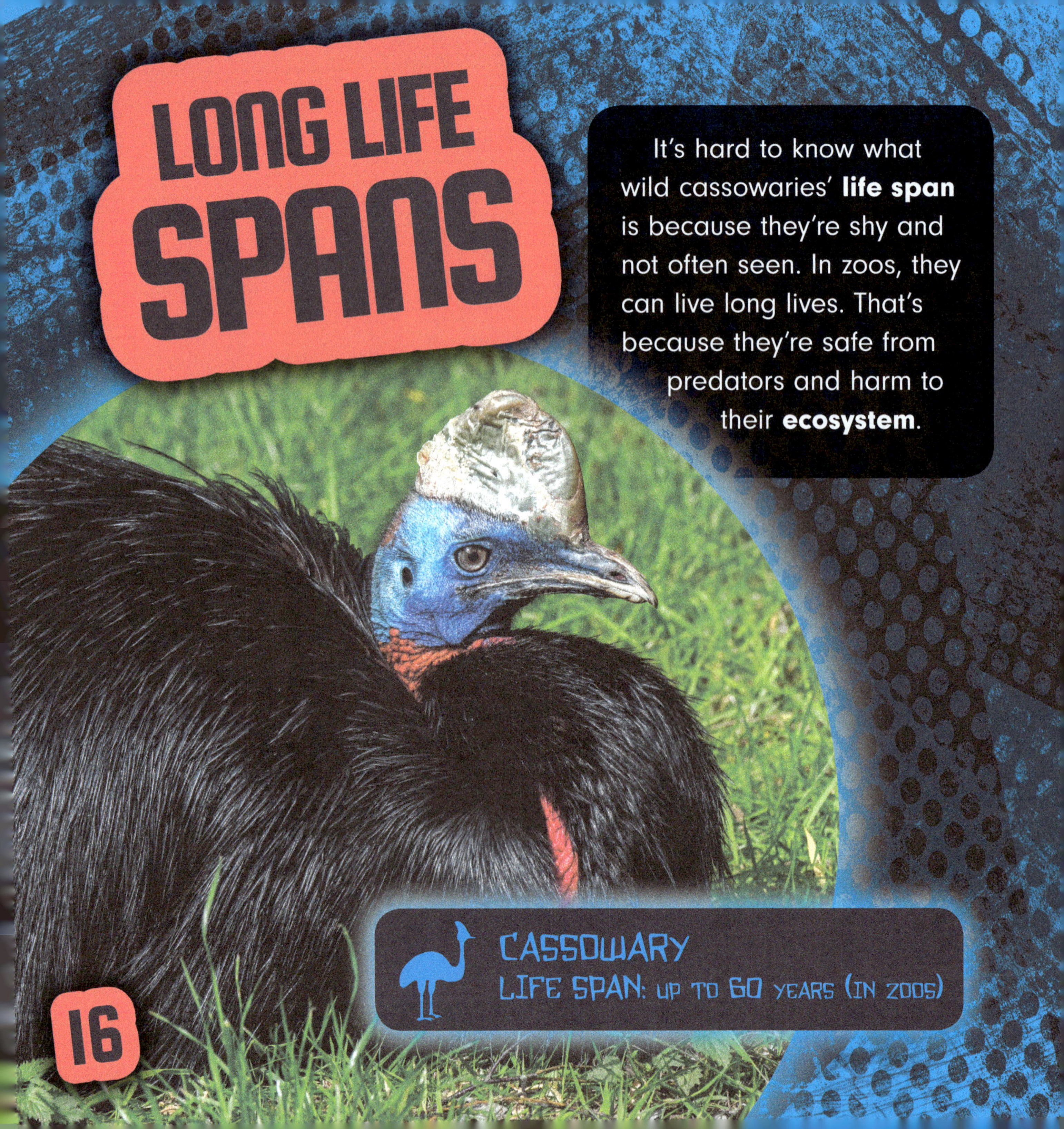

CASSOWARY
LIFE SPAN: UP TO 60 YEARS (IN ZOOS)

RHINOCEROS

LIFE SPAN: UP TO 40 YEARS (IN THE WILD)

Rhinos can live a long time. But sadly they often don't. People kill them for their horns, which are wrongly believed to treat sicknesses.

Northern cassowaries in zoos can live longer than wild rhinos. They win this match!

ECOSYSTEM SHAPERS

This next matchup is different. Which animal is most helpful to its ecosystem? Cassowaries are called rainforest gardeners because they spread seeds of many kinds of fruit in their poop. Their rainforest couldn't survive without them.

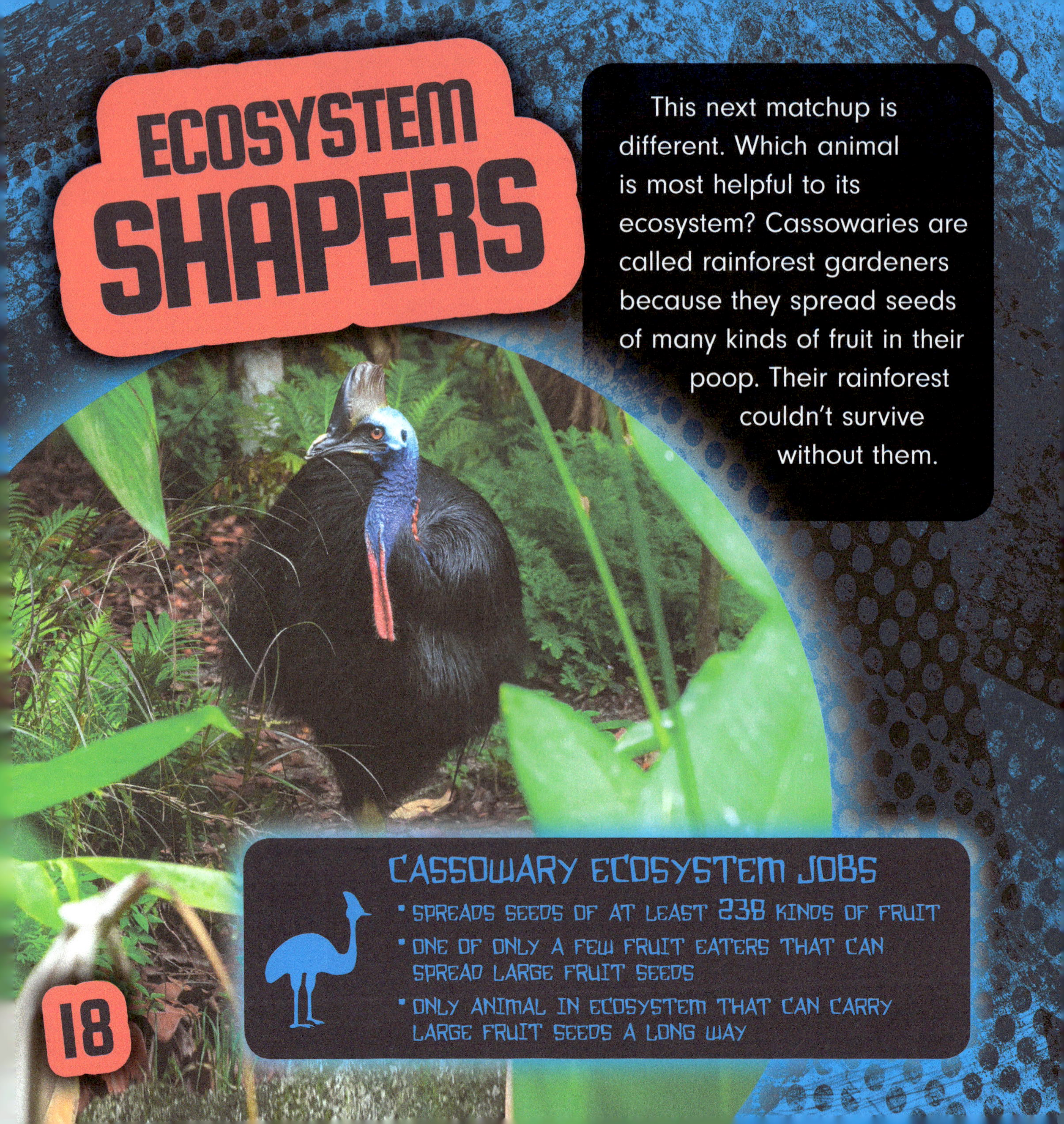

CASSOWARY ECOSYSTEM JOBS

- SPREADS SEEDS OF AT LEAST 238 KINDS OF FRUIT
- ONE OF ONLY A FEW FRUIT EATERS THAT CAN SPREAD LARGE FRUIT SEEDS
- ONLY ANIMAL IN ECOSYSTEM THAT CAN CARRY LARGE FRUIT SEEDS A LONG WAY

RHINOCEROS ECOSYSTEM JOBS

- HELPS CREATE AREAS WITH DIFFERENT KINDS OF PLANTS THAT FEED MANY KINDS OF ANIMALS
- FEEDING AREAS FIGHT CLIMATE CHANGE BY STORING THE GAS CARBON DIOXIDE

As rhinos feed, they eat just certain grasses. This allows other kinds of plants to grow, which is helpful for many plant-eating animals. These grasslands help fight **climate change** too!

Both animals are very important to their ecosystems. This looks like a tie!

WHO WINS?

Now you've learned about these two animals. You know not to get too close to either! Rhinos are fast and very heavy. They also have some scary weapons. But cassowaries can break bones with their kicks and cut enemies with their claws. Which do you think would win if they were actually to fight?

These two animals are pretty evenly matched. They're also both important to their ecosystems. We need rhinos and cassowaries in our world!

THIS BIZARRE BEAST BATTLE COULD NEVER TRULY HAPPEN. CASSOWARIES AND RHINOS LIVE IN DIFFERENT PARTS OF THE WORLD. BUT IT'S FUN TO IMAGINE WHAT MIGHT HAPPEN!

GLOSSARY

climate change: long-term change in Earth's climate, caused partly by human activities such as burning oil and natural gas

communicate: to share ideas and feelings through sounds and motions

ecosystem: all the living things in an area

frequency: how many times a sound wave is repeated in a period of time. A low-frequency sound wave produces a low-pitched sound, while a high-frequency sound wave produces a high-pitched sound.

hatch: to break open or come out of

life span: the amount of time a person or animal lives

mammal: a warm-blooded animal that has a backbone and hair, breathes air, and feeds milk to its young

mate: one of two animals that come together to produce babies

rainforest: a forest that gets lots of rain

threatened: a sense that one is in danger

ultraviolet: a range of wavelengths in light beyond the violet end of the visible color series

wattle: loose skin hanging from the neck or head of a bird

weapon: something used to fight an enemy

FOR MORE INFORMATION

BOOKS

Hansen, Grace. *Rhinoceros.* Minneapolis, MN: Abdo Kids, 2018.

Rheinstrom, David. *Animal Attack! A Wildlife Survival Guide.* New York, NY: Tangerine Press, 2019.

WEBSITES

Cassowaries
www.kidcyber.com.au/cassowaries
Learn more about cassowaries, and see some interesting photos.

Rhinoceroses
www.ducksters.com/animals/rhinoceros.php
Find out more about rhinoceroses.

Rhino Facts!
www.natgeokids.com/uk/discover/animals/general-animals/rhinoceros-facts/
Discover some amazing rhinoceros facts on this website.

INDEX